Principles and Practices for a Federal Statistical Agency

SECOND EDITION

Committee on National Statistics
Margaret E. Martin, Miron L. Straf, and Constance F. Citro, *editors*

Commission on Behavioral and Social Sciences and Education
National Research Council

NATIONAL ACADEMY PRESS
Washington, D.C.

NATIONAL ACADEMY PRESS 2101 Constitution Avenue, N.W. Washington, D.C. 20418

NOTICE: The project that is the subject of this report was approved by the Governing Board of the National Research Council, whose members are drawn from the councils of the National Academy of Sciences, the National Academy of Engineering, and the Institute of Medicine. The members of the committee responsible for the report were chosen for their special competences and with regard for appropriate balance.

The study was supported by Contract/Grant No. SBR-9709489 between the National Academy of Sciences and the National Science Foundation. Any opinions, findings, conclusions, or recommendations expressed in this publication are those of the author(s) and do not necessarily reflect the view of the organizations or agencies that provided support for this project.

International Standard Book Number 0-309-07314-6

Additional copies of this report are available from National Academy Press, 2101 Constitution Avenue, N.W., Lockbox 285, Washington, D.C. 20055
Call (800) 624-6242 or (202) 334-3313 (in the Washington metropolitan area)

This report is also available online at http://www.nap.edu

Printed in the United States of America

Suggested citation: National Research Council (2001). *Principles and Practices for a Federal Statistical Agency, Second Edition.* Committee on National Statistics. Margaret E. Martin, Miron L. Straf, and Constance F. Citro, Editors. Commission on Behavioral and Social Sciences and Education. Washington, DC: National Academy Press.

THE NATIONAL ACADEMIES

National Academy of Sciences
National Academy of Engineering
Institute of Medicine
National Research Council

The **National Academy of Sciences** is a private, nonprofit, self-perpetuating society of distinguished scholars engaged in scientific and engineering research, dedicated to the furtherance of science and technology and to their use for the general welfare. Upon the authority of the charter granted to it by the Congress in 1863, the Academy has a mandate that requires it to advise the federal government on scientific and technical matters. Dr. Bruce M. Alberts is president of the National Academy of Sciences.

The **National Academy of Engineering** was established in 1964, under the charter of the National Academy of Sciences, as a parallel organization of outstanding engineers. It is autonomous in its administration and in the selection of its members, sharing with the National Academy of Sciences the responsibility for advising the federal government. The National Academy of Engineering also sponsors engineering programs aimed at meeting national needs, encourages education and research, and recognizes the superior achievements of engineers. Dr. William A. Wulf is president of the National Academy of Engineering.

The **Institute of Medicine** was established in 1970 by the National Academy of Sciences to secure the services of eminent members of appropriate professions in the examination of policy matters pertaining to the health of the public. The Institute acts under the responsibility given to the National Academy of Sciences by its congressional charter to be an adviser to the federal government and, upon its own initiative, to identify issues of medical care, research, and education. Dr. Kenneth I. Shine is president of the Institute of Medicine.

The **National Research Council** was organized by the National Academy of Sciences in 1916 to associate the broad community of science and technology with the Academy's purposes of furthering knowledge and advising the federal government. Functioning in accordance with general policies determined by the Academy, the Council has become the principal operating agency of both the National Academy of Sciences and the National Academy of Engineering in providing services to the government, the public, and the scientific and engineering communities. The Council is administered jointly by both Academies and the Institute of Medicine. Dr. Bruce M. Alberts and Dr. William A. Wulf are chairman and vice chairman, respectively, of the National Research Council.

COMMITTEE ON NATIONAL STATISTICS
2000-2001

JOHN E. ROLPH (*Chair*), Marshall School of Business, University of Southern California
JOSEPH G. ALTONJI, Institute of Policy Research, Northwestern University
LAWRENCE D. BROWN, Department of Statistics, University of Pennsylvania
JULIE DAVANZO, RAND, Santa Monica, California
WILLIAM F. EDDY, Department of Statistics, Carnegie Mellon University
ROBERT M. GROVES, University of Michigan and Joint Program in Survey Methodology, University of Maryland
HERMANN HABERMANN, Statistics Division, United Nations, New York
JOEL L. HOROWITZ, Department of Economics, University of Iowa
WILLIAM KALSBEEK, Department of Biostatistics, University of North Carolina
RODERICK J.A. LITTLE, School of Public Health, University of Michigan
THOMAS A. LOUIS, RAND, Washington, DC
CHARLES F. MANSKI, Department of Economics, Northwestern University
EDWARD B. PERRIN, Department of Health Services, University of Washington
DARYL PREGIBON, AT&T Laboratories-Research, Florham Park, New Jersey
FRANCISCO J. SAMANIEGO, Division of Statistics, University of California, Davis
RICHARD SCHMALENSEE, Sloan School of Management, Massachusetts Institute of Technology
MATTHEW D. SHAPIRO, Department of Economics, University of Michigan

ANDREW A. WHITE, *Director*, Committee on National Statistics
MARGARET E. MARTIN, *Consultant*
MIRON L. STRAF, *Deputy Director*, Commission on Behavioral and Social Sciences and Education

CONSTANCE F. CITRO, *Senior Study Director*, Committee on National Statistics
CARRIE L. MUNTEAN, *Research Assistant*, Committee on National Statistics

http://www.nationalacademies.org/cnstat

Acknowledgments

The Committee on National Statistics thanks the many people who contributed their time and expertise to the preparation of this report. We are most appreciative of their cooperation and assistance.

In expressing our gratitude to the staff, a special measure of recognition is due to Margaret Martin and Miron Straf, who were coeditors of the original edition of this report. In revising the report, they were joined as editors of the second edition by Constance Citro. Linda Ingram and Flossie Wolf assisted the committee in preparing the original report, and Carrie Muntean, Andrew White, and Thomas Jabine assisted in preparing this second edition. Eugenia Grohman offered comments and guidance in editing the first edition. This second edition benefited from the editing of Christine McShane. We also are indebted to many others who offered valuable comments and suggestions.

This report has been reviewed in draft form by individuals chosen for their diverse perspectives and technical expertise, in accordance with procedures approved by the Report Review Committee of the National Research Council (NRC). The purpose of this independent review is to provide candid and critical comments that will assist the institution in making the published report as sound as possible and to ensure that the report meets institutional standards for objectivity, evidence, and responsiveness to the study charge. The review comments and draft manuscript remain confidential to protect the integrity of the deliberative process.

We thank the following individuals for their participation in the review of this report: Mary Ann Baily, Hastings Center, Garrison, NY; Vincent P. Barabba, General Motors Corporation, Detroit, MI; Ivan Fellegi, Statistics Canada, Ottawa, Canada; Joel Greenhouse, Department of Statistics, Carnegie Mellon University; Jay Hakes, Jimmy Carter Library, Atlanta, GA; Janet L. Norwood, Chevy Chase, MD; Alice Robbin, School of Library and Information Science, Indiana University; and Stephen Stigler, Department of Statistics, University of Chicago.

Although the reviewers listed above have provided many constructive comments and suggestions, they were not asked to endorse the conclusions or recommendations nor did they see the final draft of the report before its release. The review of this report was overseen by John T. Bailar, Professor Emeritus, Department of Health Studies, University of Chicago. Appointed by the NRC's Report Review Committee, he was responsible for making certain that an independent examination of this report was carried out in accordance with institutional procedures and that all review comments were carefully considered. Responsibility for the final content of this report rests entirely with the authoring committee and the institution.

Finally, we recognize the many federal agencies that support the Committee on National Statistics directly and through a grant from the National Science Foundation. Without their support and their commitment to improving the national statistical system, the committee work that is the basis of this report would not have been possible.

Contents

Preface to the Second Edition

In 1992 the Committee on National Statistics (CNSTAT) issued a white paper on principles and practices for a federal statistical agency. The paper responded to requests from Congress and others for advice on what constitutes an effective statistical agency. It identified and commented on three basic principles: relevance to policy issues, credibility among data users, and trust among data providers. It also discussed 11 important practices, including a strong measure of independence and commitment to quality and professional practice (National Research Council, 1992).

The CNSTAT report has been used by federal statistical agencies to inform department officials, advisory committees, and others. It has also been used in a congressionally mandated study by the U.S. General Accounting Office (1995) to evaluate the performance of major statistical agencies and in a review of the federal statistical system by a former commissioner of the Bureau of Labor Statistics (Norwood, 1995). Its principles informed the establishment and later assessment of a new statistical agency, the Bureau of Transportation Statistics (see National Research Council, 1997b).

Eight years have passed since the white paper was first issued, and the committee decided that it would be useful to release a revised and updated version at this time. This second edition does not change the basic *principles* for federal statistical agencies, because the committee believes these principles are and will continue to be important guides for effective practice. The second edition does revise and expand the discussion of some of the *practices* that characterize an effective federal statistical agency and brings the discussion up to date with references to recent reports by the committee and others.

Driving the revisions is our recognition of the need for statistical agencies to keep up to date and to meet the challenges for their missions that are posed by such technological, social, and economic changes as the widespread use of the Internet for the dissemination and, increasingly, the collection of data, the heightened concern about safeguards for confidential information, and the information requirements of a changing economy. New and revised text addresses the reasons for establishing a federal statistical agency, the necessity for and characteristics of independence of a federal statistical agency, the need for continual development of more useful data, for example, by integrating data from multiple sources, practices for fair treatment of data providers, the role of the Internet in the release of data, and the need for effective coordination and cooperation among statistical agencies to ensure that policy makers and citizens receive data that are accurate, relevant, and timely for their needs.

We stress that the principles and practices for a federal statistical agency articulated here are guidelines, not prescriptions. We intend them to be helpful not only to the agencies, from whose experience we benefited in preparing this revised edition, but also to inform others of the characteristics of effective statistical agencies that can serve policy makers in the executive and legislative branches, other data users, and the public well.

John E. Rolph, Chair
Committee on National Statistics, 2001

Preface to the First Edition

From time to time the Committee on National Statistics (CNSTAT) is asked for advice on what constitutes an effective federal statistical agency. For example, congressional staff raised the question as they were formulating legislation for a Bureau of Environmental Statistics, and the Secretary of the U.S. Department of Transportation asked CNSTAT for advice on establishing a new Bureau of Transportation Statistics, called for in the Intermodal Surface Transportation Efficiency Act of 1991. The National Research Council's Transportation Research Board had earlier turned to CNSTAT for information on common elements of the organization and responsibilities of federal statistical agencies for its study on strategic transportation data needs. Of interest in all of these requests are the fundamental characteristics that define a statistical agency and its operation.

Statistical agencies sometimes face situations that tax acceptable standards for professional behavior. Examples occur when policy makers, regulators, or enforcement officials seek access to data on individual respondents from a statistics agency or when policy interpretations are added to press releases announcing statistical data. Because the federal statistical system is highly decentralized, statistical agencies must operate under the policies and guidance of officials in many departments of government. Not all of these officials are knowledgeable about what is generally accepted as proper for a federal statistical agency, and issues involving judgments about conflicting objectives also arise.

In response to these situations, CNSTAT has prepared this "white paper" on principles and practices for a federal statistical agency. This paper brings together conclusions and recommendations made in many CNSTAT reports on specific agencies, programs, and topics, and it includes a discussion of what is meant by independence of a federal statistical agency and of the roles of research and analysis in a statistical agency. The commentary section contains supplementary information to further explain or illustrate the principles and practices.

In preparing this paper, CNSTAT and its staff solicited suggestions from many involved with federal statistical agencies. A draft of the paper was discussed by the heads of some federal statistical agencies at an open meeting of CNSTAT, and a draft was also discussed at a meeting of the Council of Professional Associations on Federal Statistics. The committee is grateful for the many suggestions and comments it received. When the report is published, CNSTAT plans to seek an even wider discussion of it at meetings of professional societies and to encourage reviews and commentaries. We hope that, in this way, the paper may evolve further and possibly influence legislation, regulations, and standards affecting federal statistical agencies.

As we were completing our work on this report, the Conference of European Statisticians drafted a resolution on the fundamental principles of official statistics in the region of the Economic Commission for Europe (ECE). Although the two documents were done independently, there is a large amount of agreement between them. We note particularly the emphasis the ECE resolution places on the need for independence for official statistics agencies (United Nations Statistical Commission and Economic Commission for Europe, 1991).[1]

Although focused on federal statistical agencies, many of the principles and practices presented here also apply to statistical activities elsewhere, particularly to those in state and local government agencies and other statistical organizations. In addition, this paper and the ECE resolution may be useful to emerging democracies that seek to establish statistical organizations in their governments.

The principles and practices articulated here are statements of best

[1]The ECE resolution was later adopted by the Statistical Commission of the United Nations (United Nations Statistical Commission, 1994). See Appendix A.

practice rather than legal or scientific rules. They are based on experience rather than law or experiment. Some of them may need to be changed as laws change, society changes, or the practice of statistics changes. They are thus intended as guidelines, not prescriptions.

Burton H. Singer, Chair
Committee on National Statistics, 1992

Part I:
Principles and Practices for a Federal Statistical Agency

Definition of a Federal Statistical Agency

Establishment of a Federal Statistical Agency

Principles for a Federal Statistical Agency

- Relevance to Policy Issues
- Credibility Among Data Users
- Trust Among Data Providers

Practices for a Federal Statistical Agency

- A Clearly Defined and Well-Accepted Mission
- A Strong Position of Independence
- Continual Development of More Useful Data
- Openness About the Data Provided
- Wide Dissemination of Data
- Cooperation with Data Users
- Fair Treatment of Data Providers
- Commitment to Quality and Professional Standards of Practice
- An Active Research Program
- Professional Advancement of Staff
- Coordination and Cooperation with Other Statistical Agencies

Note: Part I is a summary statement of principles and practices for an effective statistical agency. Part II, "Commentary," further explains, defines, and illustrates the topics in Part I.

DEFINITION OF A FEDERAL STATISTICAL AGENCY

A federal statistical agency is a unit of the federal government whose principal function is the compilation and analysis of data and the dissemination of information for statistical purposes.

The theory and methods of the discipline of statistics and related fields and the practice of the profession of statistics are brought to bear on the compilation of data, on producing information from the data, and on disseminating that information.

- The *unit* must be generally recognized as a distinct entity. It may be located within either a cabinet department or an independent agency, or it could itself be an independent agency.
- *Compilation* may include direct collection of data from individuals, organizations, or establishments or the use of administrative records. It may include assembling information from a variety of sources, including other statistical agencies, in order to produce an integrated data series, such as the national income and product accounts.
- *Analysis* may take various forms, such as developing indicators, modeling, making projections, and interpreting data. It may include research and explanations of relationships between survey statistics at various levels of aggregation and other variables. Analysis by a statistical agency does not advocate policies or take partisan positions.
- *Dissemination* means making information available to the public, to others in the executive branch, and to Congress.
- *Statistical purposes* include description, evaluation, analysis, inference, and research. For these purposes, a statistical agency may collect data from individuals, establishments, or other organizations directly, or it may obtain data from administrative records, but it does not do so for administrative, regulatory, or enforcement purposes. Statistical purposes exclude any interest in an individual person or economic unit. The data are used solely to describe and analyze statistical patterns, trends, and relationships involving groups of persons or other units.

ESTABLISHMENT OF A FEDERAL STATISTICAL AGENCY

Statistics that are publicly available from government agencies are essential for a nation to advance the economic well-being and quality of life of its people. Its public policy makers are best served by statistics that are accurate, timely, and relevant for policy decisions. Even more, the operation of a democratic system of government depends on the unhindered flow of statistical information that citizens can use to assess government actions and for other purposes. Federal statistical agencies are established to be a credible source of useful, accurate statistics in one or more subject areas that are available to the public and policy makers on a timely basis.

Accuracy requires proper concern for consistency across geographic areas and across time, as well as statistical measures of errors in the data. Timeliness requires concern for issuing data as frequently as is needed to reflect important changes in what is being studied, as well as disseminating data as soon as practicable after they are collected. Relevance requires concern for providing data that help users meet their current needs for decision making and analysis, as well as anticipating future data needs. And credibility requires concern for both the reality and appearance of impartiality, and of independence from political control. It is the primary mission of agencies in the federal statistical system to work to ensure the goals of accuracy, timeliness, relevance, and credibility of statistical information.

One reason to establish a separate statistical agency is the need for data series to be independent of control by policy makers or regulatory or enforcement agencies. Other reasons include:

- the need for information in a subject area that is updated on a continuing basis and that extends beyond the scope of individual operating units, possibly involving other departments or agencies,
- the need to protect the confidentiality of responses, and
- the opportunity to achieve greater efficiency or higher quality through a consolidated and more highly professional activity.

The principles and practices for a federal statistical agency that are reviewed in this report pertain to individual agencies as separate organizational entities in the context of a decentralized system for providing federal statistics. Historically, the response of the U.S. government to new information needs has been to create separate statistical units, so that the United States now has one of the most decentralized statistical systems of any modern nation. This report does not comment on the advantages or disadvantages of the U.S. system compared with other models for organizing government statistics. It discusses the need for federal statistical agencies to coordinate their work and cooperate with other agencies on a range of activities, describes the coordinating role of the Office of Management and Budget, and reviews some mechanisms for interagency collaboration.

PRINCIPLES FOR A FEDERAL STATISTICAL AGENCY

Relevance to Policy Issues

A federal statistical agency must be in a position to provide information relevant to issues of public policy.

A statistical agency must be knowledgeable about the issues and requirements of public policy and federal programs and able to provide objective information that is relevant to policy and program needs. In establishing priorities for statistical programs for this purpose, a statistical agency must work closely with the users of such information in the executive branch, the Congress, and interested nongovernmental groups.

Often, the provision of statistics concerning a particular subject area is itself a public policy, with the goal of serving a broad range of information needs of private- and public-sector users as well as the public. To establish priorities for such statistics, a statistical agency must maintain contact with a broad spectrum of users in the business sector, academia, state and local governments, and elsewhere.

Credibility Among Data Users

A federal statistical agency must have a relationship of mutual respect and trust with those who use its data and information.

It is essential that a statistical agency strive to maintain credibility for itself and for its data. Few data users are in a position to verify the completeness and accuracy of statistical information; they must rely on an agency's reputation as a credible source of accurate and useful statistics.

To have credibility, an agency must be and must be perceived to be free of political interference and policy advocacy. Also important for credibility is that an agency follow such practices as wide dissemination of data, openness about the quality of the data provided, commitment to quality, and fair treatment of data providers.

Trust Among Data Providers

A federal statistical agency must have a relationship of mutual respect and trust with respondents who provide data and with all data subjects whose information it obtains.

Data providers must be able to rely on the word of a statistical agency. An agency achieves credibility with its providers by ensuring appropriate confidentiality of responses. Maintaining confidentiality, in particular, precludes the use of individually identifiable information for any administrative, regulatory, or enforcement purpose.

Trust with respondents is also achieved by respecting individual privacy. Such respect requires that an agency minimize to the extent possible the time and effort of respondents to provide information and fairly inform respondents of the intended uses of their information.

PRACTICES FOR A FEDERAL STATISTICAL AGENCY

The effective operation of a federal statistical agency must begin with two related elements: a clearly defined and well-accepted mission together with a strong position of independence. With these prerequisites, effective operation involves a wide range of practices: continual development of more useful data, openness about the data provided, wide dissemination of data, cooperation with data users, fair treatment of data providers, commitment to quality and professional practice, an active research program, professional advancement of staff, and coordination and cooperation with other statistical agencies.

A Clearly Defined and Well-Accepted Mission

An agency's mission should include responsibility for all elements of its programs for providing statistical information: determining sources of data, measurement methods, efficient methods of data collection and processing, and appropriate methods of analysis; and ensuring the public availability not only of the data, but also of documentation of the methods used to obtain the data and their quality. The mission should include the responsibility for assessing information needs and priorities and ways to meet those needs, which could include the establishment of a data collection program or the modification or discontinuance of an existing program.

A Strong Position of Independence

A widely acknowledged position of independence is necessary for a statistical agency to have credibility and to carry out its function to provide an unhindered flow of useful, high-quality information for the public and policy makers. Without the credibility that comes from a strong degree of independence, users may lose trust in the accuracy and objectivity of the agency's data, and data providers may become less willing to cooperate with agency requests.

In essence, a statistical agency must be distinct from those parts of the department that carry out enforcement and policy-making activities. It must be impartial and avoid even the appearance that its collection, analysis, and reporting processes might be manipulated for political purposes or that individually identifiable data might be turned over for administrative, regulatory, or enforcement purposes.

The circumstances of different agencies may govern the form that independence takes. In some cases, the legislation that establishes the agency may specify that the agency head be professionally qualified, be appointed by the President and confirmed by the Senate, serve for a specific term not coincident with that of the administration, and have direct access to the secretary of the department in which the agency is located. These organizational requirements are associated with and facilitate the ability of a statistical agency to achieve a strong position of independence and credibility, but they are neither necessary nor sufficient.

Other characteristics related to independence are that the statistical agency have:

- authority for professional decisions over the scope, content, and frequency of data compiled, analyzed, or published. Most statistical agencies have such broad authority, limited by budgetary constraints, departmental requirements, Office of Management and Budget (OMB) review, and congressional mandates.
- authority for selection and promotion of professional, technical, and operational staff.
- recognition by policy officials outside the statistical agency of its authority to release statistical information without prior clearance.
- authority for statistical agency heads and qualified staff to speak about the agency's statistics before Congress, with congressional staff, and before public bodies.

- adherence to predetermined schedules in public release of important economic or other indicator data to prevent even the appearance of manipulation of release dates for political purposes.
- maintenance of a clear distinction between the release of statistical information and policy interpretations of such information by the President, the secretary of the department, or others in the executive branch.
- dissemination policies that foster regular, frequent release of major findings from an agency's statistical programs to the public via the media, the Internet, and other means.

Continual Development of More Useful Data

Statistical agencies must continually look to improve their data systems to provide information that is accurate, timely, and relevant for changing public policy needs. They must also continually seek to improve the efficiency of their programs for collecting, analyzing, and disseminating statistical information.

Ways for an agency to achieve these goals include:

- seeking opportunities to integrate data from more than one survey, or to integrate data from surveys with data from administrative records, with appropriate safeguards for confidentiality. When separate data sets are collected and analyzed in such a manner that they may be used together, the value of the resulting information and the efficiency of obtaining it can be greatly enhanced.
- sharing technical information and ideas with other statistical agencies. Such sharing can stimulate the development of innovative data collection, analysis, and dissemination methods that improve the accuracy and timeliness of information and the efficiency of data operations.
- establishing a balanced data collection program to provide relevant information for different types of data needs. Such a program could include one-time surveys on special topics, repeated surveys of cross-sections of the population that provide regularly updated statistics, and longitudinal surveys that track people, firms, and institutions over time and make it possible to analyze the causes and effects of changes in their circumstances.

Openness About the Data Provided

A statistical agency should be open about its data and their strengths and limitations. Data releases from a statistical program should be accompanied by a full description of the purpose of the program, the methods and assumptions used for data collection, processing, and reporting, what is known (and not known) about the quality and relevance of the data, appropriate methods for analysis that take account of variability and other sources of error in the data, and the results of research on the methods and data.

When problems are found in a previously released statistic that could affect its use, an agency should issue a correction promptly and publicly. An agency should be proactive in seeking ways to alert known and likely users of the data about the nature of the problem and the appropriate corrective action.

Wide Dissemination of Data

A statistical agency should strive for the widest possible dissemination of the data it compiles. Data dissemination should be timely and public. Also, measures should be taken to ensure that data are preserved and accessible to the public for use in future years.

Elements of an effective dissemination program include:

- an established publications policy that describes, for a data collection program, the types of reports and other data releases to be made available, the audience to be served, and the frequency of release.
- a variety of avenues for data dissemination, chosen to reach as broad a public as reasonably possible. Channels of dissemination include, but are not limited to, an agency's Internet web site, government depository libraries, and the media for regular communication of major findings.
- release of data in a variety of formats (e.g., printed reports, various kinds of computer-readable data files), so that the information can be accessed by users with varying skills and needs for data retrieval and analysis.
- procedures for release of information that preclude actual or perceived political interference. In particular, the timing of the public release of data should be the responsibility of the statistical agency. As noted earlier, adherence to predetermined release schedules for important economic

or other indicator data serves to prevent even the appearance of manipulation of release dates for political purposes.

- policies for the preservation of data that guide what data to retain and how they are to be archived for future secondary analysis.

Cooperation with Data Users

A statistical agency should consult with a broad spectrum of users of its data in order to make its products more useful. It should:

- seek advice on data concepts, methods, and products from data users as well as from other professional and technical subject-matter and methodological experts, using a variety of formal and informal means of communication that are appropriate to the types of input sought.
- seek advice from external groups, including those with relevant subject-matter and technical expertise, on its statistical program as a whole, on setting statistical priorities, and on the statistical methodologies it uses.
- endeavor to meet the needs for access to data while maintaining appropriate safeguards for the confidentiality of individual responses.
- provide equal access to data to all users.

Fair Treatment of Data Providers

To maintain credibility and a relationship of respect and trust with data subjects and other data providers, a statistical agency must observe fair information practices. Such practices include:

- policies and procedures to maintain the confidentiality of data, whether collected directly or obtained from administrative record sources.
- policies and procedures to respect the privacy of respondents, which include informing them of the purposes of data collection and the anticipated uses of the information, whether their participation is mandatory or voluntary, and, if voluntary, using appropriate informed consent procedures to obtain their information.
- further respecting privacy by minimizing the contribution of time and effort asked of respondents, consistent with the purposes of the data collection activity.
- recognizing the value of respondents' participation in data collec-

tion programs by accurately representing the statistical information they provide and by making it widely available.

- seeking the advice of respondents, as well as others, in planning the scope and content of the agency's statistical programs, designing its data collection procedures, and determining its data products.

Commitment to Quality and Professional Standards of Practice

A statistical agency should:

- use modern statistical theory and sound statistical practice in all technical work.
- develop strong staff expertise in the disciplines relevant to its mission, in the theory and practice of statistics, and in data collection, processing, analysis, and dissemination techniques.
- develop an understanding of the validity and accuracy of its data and convey the resulting measures of uncertainty to users.
- undertake ongoing quality assurance programs to improve data validity and reliability and to improve the processes of compiling, editing, and analyzing data.
- develop a strong and continuous relationship with appropriate professional organizations in the fields of statistics and relevant subject-matter areas.
- follow good practice, in reports and other data releases, in documenting concepts, definitions, data collection methodology, and measures of uncertainty, and in discussing possible sources of error.

An Active Research Program

An effective statistical agency should have a research program that is integral to its activities. Because smaller agencies may not be able to afford as extensive a research program as larger agencies, ways should be sought to foster sharing of research results and methods among agencies. Agencies can also augment their staff resources for research by obtaining the services of experts not on the agency's staff through consulting or other arrangements as appropriate.

The research program of a statistical agency should include:

- research on the substantive issues for which the data were compiled. Such research should be conducted not only to provide useful objective analytical results, but also as a means to identify potential improvements to the content of the data, suggest improvements in the design and operation of the data collection, and provide fuller understanding of the limitations of the data.
- research to evaluate and improve statistical methodology, in particular the identification and creation of new statistical measures and the development of improved methods for analyzing errors in data that are due not only to sampling variability, but also to other sources. Research should also be conducted on ways to reduce the time and effort requested of respondents and to improve the timeliness, accuracy, and efficiency of data collection, analysis, and dissemination procedures.
- research to understand the policy uses of the agency's information, in order to make the data more relevant to policy concerns and more useful for policy analysis and decision making.

Professional Advancement of Staff

A statistical agency's professional staff should be committed to the highest standards of quality work and professional practice. They should also be committed to the highest standards of professional ethics with regard to maintaining the agency's credibility as an objective, independent source of accurate and useful information obtained through fair information practices.

To develop a high-caliber staff, a statistical agency must recruit and retain qualified statisticians, analysts in fields relevant to its mission, and other staff with skills that are needed for its efficient and effective operation. An agency's personnel policies should encourage staff to maintain and extend their technical capabilities through appropriate professional and developmental activities, such as attendance at professional meetings, participation in relevant training programs, and rotation of assignments. An agency should also seek opportunities to reinforce the commitment of its staff to ethical standards of practice.

Coordination and Cooperation with Other Statistical Agencies

A statistical agency must seek opportunities to cooperate with other statistical agencies to enhance the value of its own information and that of other agencies in the federal statistical system. Although agencies differ in their subject-matter focus, there is overlap in their missions and a common interest in serving the public need for credible, high-quality statistics gathered as efficiently as possible.

When possible and appropriate, federal statistical agencies should cooperate not only with each other, but also with state and local statistical agencies in the provision of data for subnational areas. Federal statistical agencies should also cooperate with foreign and international statistical agencies to exchange information on both data and methods and to develop common classifications and procedures to promote international comparability of information.

Such cooperative activities as integrating data compiled by different statistical agencies invariably require much effort to overcome differences in agency missions and operations. But the rewards are data more relevant to policy concerns and a stronger statistical system as a whole. For these reasons, statistical agencies must act as partners to one another, not only in the development of data, but also for the entire panoply of statistical activities, including definitions, concepts, measurement methods, analytical tools, professional practice, dissemination modes, means to protect the confidentiality of responses, and ways to advance the effective use of statistical information.

Part II

Commentary

This part of the report comments on most of the topics in the principles and practices; the comments are offered to explain, illustrate, or further define the statement of principle in Part I.

DEFINITION OF A FEDERAL STATISTICAL AGENCY

A federal statistical agency is a unit of the federal government whose principal function is the compilation and analysis of data and the dissemination of information for statistical purposes.

A statistical agency may be labeled a bureau, center, division, or office or similar title, so long as it is recognized as a distinct entity. Statistical agencies have been established for several reasons: (1) to develop new information for an area of public concern (e.g., the Bureau of Labor Statistics, the National Center for Health Statistics); (2) to conduct large statistical collection and dissemination operations (e.g., the Bureau of the Census); (3) to compile and analyze statistics from sets of administrative records for policy purposes and public use (e.g., units in the Internal Revenue Service and the Social Security Administration); and (4) to develop broad and consistent estimates from a variety of statistical and administrative sources in accordance with a prespecified conceptual framework (e.g., the Bureau of Economic Analysis in the Department of Commerce and the Economic

Research Service in the Department of Agriculture). Once established, many statistical agencies engage in all these functions to varying degrees.

This definition of a federal statistical agency does not include many statistical activities of the federal government because they are not performed by distinct units, or because they do not result in the dissemination of statistics to others—for example, statistics compiled by the Postal Service to set rates or by the Department of Defense to test weapons (see, e.g., National Research Council, 1998b, on statistics and testing for defense acquisition). Nor does it include agencies whose primary functions are the conduct or support of problem-oriented research, although much of the research may be based on information gathered by statistical means, for example, by the Department of Energy's national laboratories and by the National Institutes of Health.

Finally, this definition of a statistical agency does not usually include agencies whose primary function is policy analysis and planning (e.g., the Office of Tax Analysis in the Department of the Treasury, the Office of the Assistant Secretary for Planning and Evaluation in the Department of Health and Human Services). Such agencies may collect and analyze statistical information, and statistical agencies, in turn, may perform some policy-related analysis functions for their department. However, to maintain credibility as an objective source of accurate, useful information, statistical agencies must be separate from units that are involved in developing policy and assessing policy alternatives.

The work of federal statistical agencies is coordinated through the Interagency Council on Statistical Policy (ICSP), created by the 1995 reauthorization of the Paperwork Reduction Act. The ICSP currently includes representation from 10 principal statistical agencies and from the statistical units in the Environmental Protection Agency, the Internal Revenue Service, the National Science Foundation, and the Social Security Administration (see Box II-1).

Throughout the federal government, the Office of Management and Budget recognizes more than 70 units and agencies that are not statistical agencies but that have annual budgets of $500,000 or more for statistical activities (U.S. Office of Management and Budget, 2000). Many of the considerations in the principles and practices presented here may be pertinent to these agencies. Similarly, the principles and practices may be relevant to statistical units in state and local government agencies, and international audiences may find them useful as well.

BOX II-1
Federal Agencies Represented on the Interagency Council on Statistical Policy as of 2000

Office of Management and Budget (OMB), Chair
Bureau of the Census, Department of Commerce
Bureau of Economic Analysis (BEA), Department of Commerce
Bureau of Justice Statistics (BJS), Department of Justice
Bureau of Labor Statistics (BLS), Department of Labor
Bureau of Transportation Statistics (BTS), Department of Transportation
Economic Research Service (ERS), Department of Agriculture
Energy Information Administration (EIA), Department of Energy
National Agricultural Statistics Service (NASS), Department of Agriculture
National Center for Education Statistics (NCES), Department of Education
National Center for Health Statistics (NCHS), Department of Health and Human Services
Office of Environmental Information, Environmental Protection Agency (EPA)
Office of Research, Evaluation, and Statistics, Social Security Administration (SSA)
Science Resources Studies Division, National Science Foundation (NSF)
Statistics of Income Division (SOI), Internal Revenue Service

ESTABLISHMENT OF A FEDERAL STATISTICAL AGENCY

One of the most important reasons for establishing a statistical agency is to provide information that will allow for an informed citizenry. A democracy depends on an informed electorate. A citizen has a right to information that is relevant, accurate, and timely. Timely information of high quality is also critical to policy analysts and decision makers in both the public and private sectors. (For more information on the purposes of official statistics, see the Fundamental Principles of Official Statistics of the United Nations Statistical Commission in Appendix A.) Federal statistical agencies serve the key functions of providing a broad array of information to the public and policy makers and of ensuring the necessary quality and credibility of the data.

Private-sector organizations also provide useful statistical information, including data they compile and data collected by government agencies and others to which they add value. However, because the benefits of statistical information are shared widely throughout society and because it is difficult to collect payments for these benefits, private markets are not likely to provide all of the data that are needed for public and private decision making or to make data as widely available as needed for important public purposes. Government statistical agencies are established to ensure that a broad range of information is publicly available. (See National Research Council, 1999b, for a discussion of the governmental role in providing public goods, or near-public goods, such as research and data.)

The United States collected and published statistics long before any distinct federal statistical agency was formed (see Duncan and Shelton, 1978; Norwood, 1995). The U.S. Constitution mandated the conduct of a decennial census of population beginning in 1790, and the census enumeration was originally conducted by U.S. marshals as just one of their many duties. Legislation providing for the compilation of statistics on agriculture, education, and income was enacted by Congress in the 1860s. The Bureau of Labor (forerunner of the Bureau of Labor Statistics) was established by law in 1884 as a separate agency with a general mandate to respond to widespread public demand for information on the conditions of industrial workers. The Bureau of the Census was established as a permanent agency in 1902 to conduct the decennial census and related statistical activities.

Many federal statistical agencies that can trace their roots back to the 19th or early 20th century, such as the National Center for Education Statistics and the National Center for Health Statistics, were organized in their current form following World War II. Several new agencies were also established, such as the Energy Information Administration and the Bureau of Justice Statistics. In every case, the agency itself, in consultation with users of its information, has major responsibility for determining its specific statistical programs and for setting priorities. Initially, many of these agencies also had responsibilities for certain policy analysis functions for their department heads. More recently, policy analysis has generally been located in separate units that are not themselves considered to be statistical agencies.

A statistical agency has at least two roles: (1) provider of the statistical information and analysis needed for policy and program administration by its own department and (2) source of national statistics for the public in its area of concern. It is sometimes difficult to keep these two roles distinct on policy-relevant statistics. An effective statistical agency, nevertheless, will frequently play a creative, not just reactive, role in the development of data needed for policy analysis. Sometimes federal statistical agencies play additional roles, such as monitor and consultant on statistical matters to other units within the same department (see, e.g., National Research Council, 1985a) and collector of data on a reimbursable basis for other agencies.

There is no set rule or guideline for when it is appropriate to establish a separate federal statistical agency, carry on statistical activities within the operating units of departments and independent agencies, or contract for statistical services from existing federal statistical agencies or other organizations. Establishment of a federal statistical agency may be considered when one or more of the following conditions prevail:

- There is a need for information extending beyond one-time uses and the scope of individual operating units, possibly involving other departments and agencies. Such needs may require coordinating data from various sources, initiating new data collection programs to fill gaps, or developing regularly updated time series of estimates.
- There is a need, in fact or as a matter of credibility, to ensure that major data series are independent of policy makers' control.
- There is a need to establish confidentiality of data by law or regulation covering a distinct organizational unit. When a separate statistical unit is established, the data it collects that could identify individual reporting units can be more easily protected by law or regulation from disclosure. Statistical agencies disseminate statistical data for statistical purposes; they do not disseminate identifiable data for administrative, regulatory, or enforcement uses. The functional separation of statistical data, recommended by the Privacy Protection Study Commission (1977), is easier to maintain when the data are compiled in a unit separate from operating units. At the same time, functional separation makes the promise of confidentiality more credible.
- There is a need to emphasize the principles and practices of an effective statistical agency, for example, professional practice, openness about the data provided, and wide dissemination of data.
- There is a need to encourage research and development of a broad range of statistics in a particular area of public interest or of government activity or responsibility.
- There is a need to consolidate compilation, analysis, and dissemina-

tion of statistics in one unit to encourage high-quality performance, eliminate duplication, and streamline operations.

PRINCIPLES FOR A FEDERAL STATISTICAL AGENCY

A federal statistical agency must be in a position to provide information relevant to issues of public policy.

A statistical agency supplies information not only for the use of immediate managers and policy makers in the executive branch and for legislative designers and overseers in Congress, but also to all those who require statistical information on public issues, whether the information is needed for purposes of production, trade, consumption, or participation in civic affairs. Just as a free enterprise economic system depends on the availability of economic information to all participants, a democratic political system depends on wide access to information on education, health, transportation, the economy, the environment, criminal justice, and other social concerns.

Federal statistical agencies are responsible for providing statistics on conditions in a variety of areas. The resulting information is used both inside and outside the government not only to delineate problems and sometimes to suggest courses of action, but also to evaluate the results of government activity or lack of activity. The statistics provide much of the basis on which the government itself is judged. This role places a heavy responsibility on federal statistical agencies for impartiality and objectivity.

In order to provide information that is relevant for public policy, statistical agencies need to reach out to users of the data. Federal statistical agencies usually are in touch with the primary users in their own departments. Considerable energy and initiative are required to open avenues of communication more broadly to other current and potential users, including analysts and policy makers in other federal departments, state and local government agencies, academic researchers, private-sector organizations, organized constituent groups, the media, and Congress. Advisory committees representing major users are frequently employed and are recommended as a means to obtain users' views (see, e.g., National Research Council, 1993a).

One frequently recommended method for alerting statistical agencies to emerging statistical information needs is for the agency's own staff to engage in analysis of its data (Norwood, 1975; Martin, 1981; Triplett,

1991). For example, relevant analysis may use the agency's data to examine correlates of key social or economic phenomena or to study the statistical error properties of the data. Such in-house analysis can lead to improvements in the statistics, to identification of new needs, to a reordering of priorities, and to closer cooperation and mutual understanding with policy analysis units. In its work for a policy analysis unit, a statistical agency describes conditions and possibly measures progress toward some previously identified goal, but it refrains from making policy recommendations. The distinction between statistical analysis and policy analysis is not always clear, and a statistical agency will need to consider carefully the extent of policy-related activities that are appropriate for it to undertake.

A federal statistical agency must have a relationship of mutual respect and trust with those who use its data and information.

Users of a statistical agency's data must be able to trust that the data were collected and analyzed in an objective, impartial manner and that they are as reliable as the agency can make them. An agency should make every effort to provide accurate and credible statistics that will permit policy debates to be concerned about policy, not about the credibility of the data. Credibility is enhanced when an agency fully informs users of the strengths and weaknesses of the data, makes data available widely, and consults with users about priorities for data collection and analysis.

A federal statistical agency must have a relationship of mutual respect and trust with respondents who provide data and all data subjects whose information it obtains.

The statistics program of the federal government relies in large part on information supplied by individuals and by organizations outside the federal government, such as state and local governments, businesses, and other organizations. Some of this information is required by law or regulation (such as employers' wage reports), some of it is related to administration of government programs (such as information provided by benefit recipients), but much of it is obtained through the voluntary cooperation of respondents in statistical surveys. Even when response is mandatory, the cooperation of respondents reduces costs and is likely to promote accuracy (see National Research Council, 1995). Important elements in encouraging such cooperation are that respondents believe that the data requested are important, that they are being collected in an impartial, competent manner, and that the confidentiality of their responses will be protected.

In brief, trust in a statistical agency must be maintained. The agency must not be perceived as being swayed by political considerations. It must be perceived as working in the national interest, not the interest of a particular administration, and as taking a long view, balancing new data needs against the need for consistency with past data (Ryten, 1990). Respondent trust also depends on providing respondents with realistic promises of confidentiality that the agency can reasonably expect to honor and then scrupulously honoring those promises.

PRACTICES FOR A FEDERAL STATISTICAL AGENCY

A Clearly Defined and Well-Accepted Mission

A clear understanding of the mission of an agency, the scope of its statistical programs, and its authority and responsibilities is basic to planning and evaluating its programs and to maintaining credibility and independence from political control (National Research Council, 1986). Some agency missions are clearly spelled out in legislation; other agencies have only very general legislative authority. On occasion, very specific requirements may be set by legislation or regulation.

Agencies should communicate their mission clearly to others. The use of the Internet is one means to publicize an agency's mission to a broad audience and to provide related information, including enabling legislation, the scope of the agency's statistical program, confidentiality provisions, and operating procedures.

An agency's mission should focus on the compilation, evaluation, analysis, and dissemination of statistical information. In addition, considerable and formal attention must be paid to setting statistical priorities (National Research Council, 1976). Advice from outside groups should be sought on the agency's statistical program, on setting statistical priorities, on the statistical methods used, and on data products. Such advice may be sought in a variety of formal and informal ways, and it should be obtained from data users and providers as well as professional or technical experts in the subject-matter area and in statistical methods and procedures. A strong research program in the agency's subject-matter field can assist in setting priorities and identifying ways to improve an agency's statistical programs (Triplett, 1991).

A Strong Position of Independence

A statistical agency must be able to provide credible information that may be used to evaluate the program and policies of its own department or the government as a whole. More broadly, a statistical agency must be a trustworthy source of objective, reliable information for decision makers, analysts, and others inside and outside the government who want to use statistics to understand current conditions, draw comparisons with the past, and help guide plans for the future. For these purposes, a strong position of independence for a statistical agency is essential. (See the Fundamental Principles of Official Statistics of the United Nations Statistical Commission in Appendix A.)

Statistical agency independence must be exercised in a broader framework. Legislative authority usually gives ultimate responsibility to the department rather than the statistical agency head. In addition, an agency is subject to the normal budgetary processes and to various coordinating and review functions of the Office of Management and Budget (OMB), as well as the legislative mandates, oversight, and informal guidance of Congress.

Within this broader framework, a statistical agency must work to maintain its credibility as an impartial purveyor of information. In the long run, the effectiveness of an agency depends on its maintaining a reputation for impartiality; thus, an agency must be continually alert to possible infringements on its credibility and be prepared to argue strenuously against such infringements.

Independence of an agency head is usually encouraged when the head is appointed by the President with approval by the Senate. Examples of agencies with such an arrangement are the Bureau of the Census, the Bureau of Justice Statistics, the Bureau of Labor Statistics, the Bureau of Transportation Statistics, the Energy Information Administration, and the National Center for Education Statistics. A further safeguard is provided when such a head is appointed for a fixed term, as is currently the case with the Bureau of Labor Statistics, the Bureau of Transportation Statistics, and the National Center for Education Statistics. It is desirable that the term not coincide with the presidential term, so that incumbents need not end their leadership with changes of administration and professional considerations may more easily predominate over political aims in the appointment process.

It is also desirable that a statistical agency head have direct access to the secretary of the department or the head of the independent agency in which it is located. Such access allows the head to inform new secretaries about the appropriate role of a statistical agency and present the case for new statistical initiatives to the secretary directly. Among the agency heads with presidential appointments, such direct access currently is provided by legislation only for the Bureau of Labor Statistics and the Bureau of Transportation Statistics.

These organizational aspects—appointment by the President with approval by the Senate, a fixed term not coincident with that of the administration, and direct access to the secretary of the agency's department—facilitate a strong position of independence for a statistical agency. However, they are neither necessary nor sufficient.

Control over personnel actions, especially the selection and appointment of qualified professional staff, including senior executive career staff, is another aspect of independence. Agency staff reporting directly to the agency head should have formal education and deep experience in the substantive, methodological, operational, or management issues facing the agency as appropriate for their positions. In addition, professional qualifications are of the utmost importance for statistical agency heads, whether the profession is that of statistician or the subject-matter field of the statistical agency (National Research Council, 1997b). The American Statistical Association has, when requested, assisted in the development of a list of suitable candidates for heads of statistical agencies.

Authority to decide the scope and content of the data collected or compiled is an important element of independence. Most statistical agencies have broad authority, limited by budgetary constraints, departmental interests, OMB review, and congressional mandates. In addition, the courts sometimes become involved in interpreting laws and regulations that affect statistical agencies, as in a number of issues concerning confidentiality and freedom of information, as well as in the issue of adjusting the census population counts.

Congress frequently specifies particular data that it wishes to be collected (e.g., by the National Agricultural Statistics Service in the Department of Agriculture, the National Center for Health Statistics in the Department of Health and Human Services) and, in the case of the decennial census, requires an opportunity to review the proposed questions before the forms are printed. The OMB Office of Information and Regulatory Affairs, under the Paperwork Reduction Act (and under the preceding Federal Reports Act), has the responsibility for designating a single data collection instrument for information wanted by two or more agencies. It also has the responsibility under the same act for reviewing all questionnaires and other instruments for the collection of data from 10 or more respondents.

The budgetary constraints on statistical agencies and OMB review of data collections are ongoing; the other pressures depend, in part at least, on the relations between a statistical agency and those who have supervisory or oversight functions. Agencies need to develop skills in communicating to oversight groups the need for statistical series and credibility in assessing the costs of statistical work. In turn, although it is standard practice for the secretary of a department or the head of an independent agency to have ultimate responsibility for all matters within the department or agency, for credibility, the head of a statistical agency should be allowed full authority in professional and technical matters.

Other aspects of independence that underscore a statistical agency's credibility are also important. These aspects include authority to release statistical information without prior clearance and authority for the statistical agency head and qualified staff to speak about the agency's statistics before Congress, with congressional staff, and before public bodies.

It is important, when a statistical agency releases information publicly, that a clear distinction be made between the statistical information and any policy interpretations of such information by the secretary of the department, the President, or others. Not even the appearance of manipulation for political purposes should be allowed. This is one reason why statistical agencies adhere to predetermined schedules for the public release of important economic indicators and take steps to ensure that no person outside the agency can gain access to such indicators before the official release time (see U.S. Office of Management and Budget, 1985).

Continual Development of More Useful Data

Federal statistical agencies cannot be static. To provide information of continued relevance for public and policy use, they must continually anticipate data needs for future policy considerations and look for ways to develop data systems that can serve broad purposes. To improve the quality and timeliness of their information, they must keep abreast of methodological and technical advances and be prepared to implement new procedures in a timely manner. They must also continually seek ways to make their operations more efficient. Preparing for the future requires that agencies reevaluate existing data series, plan new data series as required, and be innovative and open in their consideration of ways to improve their programs.

Because of the decentralized nature of the federal statistical system, innovation often requires cross-agency collaboration. For example, an effective way to increase the usefulness of survey data is to integrate them with data from other surveys or with data from administrative records, such as social program records. Such integration typically requires that several agencies work together. For example, in the area of health care statistics, a study by a panel of the Committee on National Statistics concluded that no single survey was likely ever to meet all the criteria, address all the technical problems, or meet all users' needs for data. In order to provide adequate information on the availability, financing, and quality of health care, a coordinated and integrated system of data collection activities involving several organizational entities was required (National Research Council and Institute of Medicine, 1992).

Innovation also implies a willingness to implement different kinds of data collection efforts to answer different needs. For example, the need to understand temporal changes in important social or economic events may call for the development of longitudinal surveys that track people, institutions, or firms over time. Statistical agencies have developed useful longitudinal surveys. However, because agencies are oriented toward the mission of their particular department, such surveys (and cross-sectional data activities as well) are typically focused on population groups (or other entities) that the department serves. So information is available on the health status of infants and young children, on the educational performance of children in schools, and on participation in the labor force by working age adults. But the health status of young children affects educational performance, and educational performance affects labor force outcomes. Longitudinal surveys that track population groups across the important transitions from early childhood to school and from school to the labor force are important to consider (National Research Council, 1998a).

Developing longitudinal data often requires much coordination with policy research agencies and with academic researchers and, especially in the case of children, requires coordination across many departments of government. Longitudinal data are often more expensive to collect than cross-sectional data and generally require more sophisticated methods for collection and analysis. In addition, more time may be needed to produce useful data products from longitudinal surveys. Yet data from longitudinal surveys are potentially very useful—sometimes they are the only means to answer important policy questions (see, e.g., National Research Council, 1997a, on data needs to inform retirement income policy).

Another area in which it is important for statistical agencies to be innovative concerns the methods used for data collection, analysis, and dissemination. Agencies need to investigate new or modified methods that have the potential to improve the accuracy and timeliness of their data and the efficiency of their operations. Careful evaluation of new methods is required to assess their benefits and costs in comparison to current methods and to determine effective implementation strategies.

For example, experience with the use of computer-assisted interviewing techniques, which many agencies have adopted for data collection, has identified benefits. It has also identified challenges for the timely provision of data and documentation that require continued research to develop solutions that maximize the gains from these techniques. Similarly, agencies need to carefully evaluate their growing use of the Internet, which has become a standard vehicle for data dissemination and is increasingly being used for data collection. Internet dissemination facilitates the timely availability of data to a broad audience and provides a valuable tool for users to learn of related data sets from other agencies. However, it poses challenges in several areas, such as how best to provide information on data quality and appropriate use of the data to an audience that spans a wide range of analytical skills and understanding.

Openness About the Data Provided

An important means to instill credibility and trust among data users and data providers is for an agency to operate in an open manner with regard to the limitations of its data. Openness requires that an agency provide a full description of its data with acknowledgment of any uncertainty and a description of the methods used and assumptions made. Agencies should provide to users reliable indications of the kinds and amounts of statistical error to which the data are subject (President's Commission on Federal Statistics, 1971). Some statistical agencies have developed detailed quality profiles for some of their major series. These have proved helpful to experienced users and agency personnel responsible for the design and operation of major surveys and data series.

Openness about data limitations requires much more than estimates of sampling error. In addition to a discussion of aspects that statisticians recognize as nonsampling errors, such as coverage errors, nonresponse, measurement errors, and processing errors, a description of the concepts used and how they relate to the major uses of the data is desirable. Descriptions of the shortcomings of and problems with the data should be provided in sufficient detail to permit the user to take them into account in the analysis and interpretation of the data.

Openness also means that a statistical agency should describe how decisions on methods and procedures were made for a data collection program. It is important to be open about research conducted on methods and data and other factors that were weighed in a decision.

Openness means as well that when mistakes are discovered after a statistic is released, the agency has an obligation to issue a correction publicly and in a timely manner. It should not only use the same dissemination vehicles to announce corrections that it used to release the original statistic, but also use additional vehicles, as appropriate, to alert the widest possible audience of current and potential users.

In summary, agencies should make an effort to provide information on the quality, limitations, and appropriate use of their data that is as frank and complete as possible. Such information, which is sometimes termed "metadata," should be made available in ways that are easy for users to access and understand, recognizing that users differ in their level of understanding of statistical data (see National Research Council, 1993a, 1997b). Agencies need to work to educate users that all data contain some uncertainty and error, which does not mean the data are wrong but that they must be used with care.

Wide Dissemination of Data

A statistical agency must have vigorous and well-planned dissemination programs to get information into the hands of users who need it on a timely basis. Planning should be undertaken from the viewpoint that the public has contributed the data elements, has paid for the data collection and processing, and should in return have access to the information in ways that make it as useful as possible to the largest number of users.

A good dissemination program provides data to users in forms that are suited to their needs. Data release may take the form of regularly updated time series, cross-tabulations of aggregate characteristics of respondents, and analytical reports that are made available in printed publications, on computer-readable media (e.g., CD-ROM), and on the Internet. (See Appendix B for a number of federal statistical agency web sites, many of which are accessible from a single source: www.fedstats.gov).

Yet another form of dissemination involves access to microdata files, which make it possible to conduct in-depth research in ways that are not possible with aggregate data. Public-use microdata files may be developed for general release. Such files contain data for individual respondents that have been processed to protect confidentiality by such means as deleting or aggregating any information that might permit individual identification. Or an agency may provide a facility on the Internet to allow users to aggregate survey data to suit their purposes, with safeguards so that the data cannot be retabulated in ways that could identify individual respondents. Or access to data may be restricted in some cases to secure sites to which researchers must come to conduct their analysis and follow stringent procedures for protecting confidentiality. Agencies should consider all forms of dissemination in order to gain the most use of their data consistent with protecting the confidentiality of responses.

A good dissemination program also uses a variety of channels to inform the broadest possible audience of potential users about available data products and how to obtain them. Such channels may include providing direct access to data on the Internet, depositing data products in libraries, establishing a network of data centers (such as the Census Bureau's state data centers), and maintaining lists of individuals and organizations to notify about new data. Agencies should also arrange for archiving of data with the National Archives and Records Administration and other data archives, as appropriate, so that data are available for historical research in future years.

An effective dissemination program provides not only the data, but also information about the strengths and weaknesses of the data in ways that can be comprehended by diverse audiences. Information about the limitations of the data should be included in every form of data release, whether in a printed report, on a computer-readable data file, or on the Internet.

On occasion, the objective of presenting the most accurate data possible may require more time than is consistent with the needs of users for the information. The tension between frequency and promptness of release on one hand and accuracy on the other should be explicitly considered. When concerns for timeliness prompt the release of preliminary estimates (as in some economic indicators), consideration should be given to the frequency of revisions and the mode of presentation of revised figures from the point of view of the users as well as the issuers of the data. Agencies that release preliminary estimates must educate the public about differences among preliminary, revised, and final estimates.

Cooperation with Data Users

Users of federal statistical data span a broad spectrum of interests and needs. They include policy makers, planners, administrators, and researchers in federal agencies, state and local governments, the business sector, and academia. They also include activists, citizens, students, and media representatives. An effective statistical agency endeavors to learn about its data users and to obtain input from them on the agency's statistical programs.

The needs of users can be explored by forming advisory committees, holding focus groups, analyzing requests and Internet activity, or by undertaking formal surveys of users. The task requires continual alertness to the changing composition and needs of users and the existence of potential users. An agency should cooperate with professional associations, institutes, universities, and scholars in the relevant fields to determine the needs of the research community and obtain their insight on potential uses. An agency should also work with relevant associations and other organizations to determine the needs of business and industry for its data.

Within the limitations of its confidentiality procedures as noted above, an agency should seek to provide maximum access to its data, including making the data available to external researchers for secondary analysis (National Research Council, 1985b). Having data accessible for a wide range of analyses increases the return on the investment in data collection and provides support for an agency's program. Once statistical data are made public, they may be used in numerous ways not originally envisaged. An agency should attempt to monitor the major uses of its data as part of its efforts to keep abreast of user needs.

Researchers and other users of data frequently request data from statistical agencies for specific purposes. The agency should have procedures in place for referring users to professionals within the agency who can comprehend the user's purposes and needs and who have a thorough knowledge of the agency's data. Statistical agencies should view these services as a part of their dissemination activities.

Ensuring equal access requires avoiding release of data to selected individuals, organizations, or news media in advance of other users. Agencies that prepare special tabulations of their data on request for external groups must be alert to the proposed uses. If the data are to be used in court cases, administrative proceedings, or collective bargaining negotiations, it is wise to have a known policy ensuring that all sides receive the special tabulations, regardless of which side requested them or paid the cost of the tabulation.

Fair Treatment of Data Providers

Protecting Confidentiality

Data providers must believe that the data they give to an agency cannot be used to harm them. For statistical data collection programs, protecting the confidentiality of individual responses is considered essential to encourage high response rates and accuracy of response. Some agencies have legislative mandates supporting promises of confidentiality; others rely on strong statements of policy, legal precedents in court cases, or custom. The latter agencies risk having their policies overturned by judicial interpretations of legislation or executive decisions that may require the agency to disclose identifiable data collected under a pledge of confidentiality. Agencies that lack strong legal protection for confidentiality should be especially careful not to give data providers stronger promises of confidentiality than they can reasonably expect to honor.

To give additional weight and stature to policies that statistical agencies have pursued for decades, OMB issued a Federal Statistical Confidentiality Order on June 27, 1997. This order assures respondents who provide statistical information to specified agencies that their responses will be held in confidence and will not be used against them in any government action, "unless otherwise compelled by law" (U.S. Office of Management and Budget, 1997).

The heads of statistical agencies must be prepared to deal with requests from other units in their own department, from other agencies and organizations, and from the courts wanting to use individually identifiable data. When such uses would be contrary to confidentiality pledges to data providers, agency heads should do everything in their power under the law to deny access to the data. In all such circumstances, agencies must be prepared to stand firm and to justify the importance of a strong commitment to confidentiality for maintaining credibility and trust with the public, in particular with data providers, and therefore in maintaining the future quality and credibility of their statistics.

Statistical agencies devote much time and effort to avoid inadvertent disclosure of confidential information in disseminating data. Recently, the widespread dissemination of statistical data via the Internet has heightened attention by agencies to effective safeguards for confidential information. Risks are increased when data for small groups are tabulated, when the same data are tabulated in a variety of ways, or when public-use microdata files (samples of records for unidentified individuals or units) are released with highly detailed geographic or other characteristics. Because of the disclosure risks associated with detailed tabulations and public-use microdata files, there is always a tension between the desire to safeguard confidentiality and the desire to provide broader public access to data. This dilemma is an important one to federal statistical agencies, and it has stimulated ongoing efforts to develop new statistical and administrative procedures to safeguard confidentiality while permitting more extensive access. An effective federal statistical agency will exercise judgment in determining which of these procedures are best suited to its requirements to serve data users while protecting confidentiality. (For discussion of these issues and alternative procedures, see the report of the Panel on Confidentiality and Data Access [National Research Council, 1993b] and the report of the Workshop on Improving Access to and Confidentiality of Research Data [National Research Council, 2000].)

Respecting Privacy

To promote trust and encourage accurate response, it is important that statistical agencies respect the privacy of respondents to the extent possible. When data providers are asked to participate in a survey, they should be told whether the survey is mandatory or voluntary, how the data will be used, and who will have access to the data. In the case of voluntary surveys, information on these matters is necessary in order for data providers to give their informed consent to participate.

Respondents invest time and effort in replying to surveys. The amount of effort varies considerably from survey to survey, depending on such factors as the complexity of the information that is requested. Statistical agencies should attempt to minimize such effort, to the extent possible, by using concepts and definitions that fit respondents' common understanding; by simplifying questionnaires; by allowing alternative modes of response (e.g., via the Internet) when appropriate; and by using administrative records or other data sources, if they are sufficiently complete and accurate to provide some or all of the needed information. In surveys of businesses or other institutions, agencies should seek innovative ways to obtain information from the institution's records and minimize the need for respondents to reprocess and reclassify information. It is also the responsibility of agencies to use qualified, well-trained interviewers. As provided in OMB directives, respondents should be informed of the likely duration of a survey interview and, if the survey involves more than one interview, how many times they will be contacted over the life of the survey. This information is particularly important when respondents are asked to cooperate in extensive interviews, search for records, or participate in longitudinal surveys.

Ways in which participation in surveys can be made easier for respondents and result in more accurate data can be explored by such means as focus group discussions or surveys. Many agencies apply the principles of cognitive psychology to questionnaire design, not only to make the resulting data more accurate, but also to make the time and effort of respondents more efficient (National Research Council, 1984). Some agencies thank respondents for their cooperation by providing them with brief summaries of the information after the survey is compiled.

A reason that respondents reply to statistical surveys is because they have been persuaded that their answers will be useful to the government or to society generally. Statistical agencies should respect this contribution by compiling the data and making them accessible to users in convenient forms. A statistical agency has an obligation to publish statistical information from the data it has collected unless it finds the results invalid.

Commitment to Quality and Professional Standards of Practice

The best guarantee of high-quality results is a strong professional staff that includes experts in the subject-matter fields covered by the agency's program, experts in statistical methods and techniques, and experts in data collection, processing, and other operations. A major function of an agency's managers is to strike a balance among these groups and promote working relationships that make the agency's program as productive as possible, with each group of experts contributing to the work of the others.

An effective statistical agency keeps up to date on developments in theory and practice that might be relevant to its program. An effective agency is also alert to changes in the economy or in society that may call for changes in the concepts or methods used in particular data sets. Often the need for change conflicts with the need for comparability with past data series, and this issue can easily dominate consideration of proposals for change. Agencies have the responsibility to manage this conflict by initiating more relevant data series while producing statistical bridges between old and new series.

An Active Research Program

Substantive Research and Analysis

There are strong arguments for a statistical agency to have staff whose responsibility is to conduct objective substantive analyses of the data that the agency compiles, such as analyses that assess trends over time or compare population groups:

- Agency analysts are in a position to understand the need for and purposes of the data and know how the statistics will be used. Such information must be available to the agency and understood thoroughly if the survey design is to produce the data required.
- Those involved in analysis can best articulate the concepts that should form the basic framework of a statistical series. Agency analysts are well situated to understand and transmit the views of external users and researchers; at the same time, close working relationships between analysts and data producers are needed for the translation of the conceptual framework into the design and operation of the survey.
- Agency analysts have access to the microdata and so are in a better position than analysts outside the agency to understand and describe the limitations of the data for analysis purposes.
- Substantive research by analysts on an agency's staff will have credibility because of the agency's commitment to openness about the data provided and maintaining independence from political control.
- Substantive research by analysts on an agency's staff can assist in formulating the agency's data program, suggesting changes in priorities, concepts, and needs for new data or discontinuance of outmoded or little-used series.

As with descriptive analyses provided by the agency, substantive analyses must be designed to be relevant to policy but not take positions on policy options or be designed with any particular policy agenda in mind.

These issues are discussed in Norwood (1975), Martin (1981), and Triplett (1991).

Methodological Research

For statistical agencies to be innovative in methods for data collection, analysis, and dissemination, methodological research must be ongoing. Such research may be directed toward improving survey design; measuring and, when possible, reducing error from such sources as nonresponse and reporting errors; making data collection, processing, and dissemination operations more efficient; reducing the time and effort asked of respondents; or developing new and improved summary measures and estimation techniques.

Much of what is current practice in statistical agencies was developed through research they conducted or obtained from other agencies. Federal statistical agencies, frequently in partnership with academic researchers, pioneered the applications of statistical probability sampling, the national economic accounts, input-output models, and other analytic methods. The U.S. Census Bureau pioneered the use of computers for processing the census, and research on data collection, processing, and dissemination operations continues to lead to creative uses of automated procedures and equipment in these areas. Several federal statistical agencies sponsor research using principles of cognitive psychology to improve the design of questionnaires, the clarity of data presentation, and the ease of use of electronic data collection and dissemination tools such as the Internet. Such research has been furthered by interactions between statistical agencies and the academic community. The history of the statistical agencies has shown repeatedly that methodological research can lead to large productivity gains in statistical activities at relatively low cost.

Research on Policy Uses

Much more needs to be known on how statistics are actually used in the policy-making process, both inside and outside the government. Research on how the information produced by a statistical agency is used in practice should contribute to future improvements in the design, concepts, and format of data products. For example, public-use files of statistical microdata were developed in response to the growing analytic needs of government and academic researchers.

Gaining an understanding of the variety of uses and users of an agency's data is only a first step. More in-depth research on the policy uses of an agency's information might, for example, explore the use of data in microsimulation or other economic models, or go further to examine how the information from such models and other sources is used in decision making (see National Research Council, 1991a, 1991b). The focus of such research should be on ways to improve the relevance and accuracy of an agency's data for use in policy analysis and decision making, independent of any particular policy agenda.

Professional Advancement of Staff

An effective federal statistical agency has personnel policies that encourage the development and retention of a strong professional staff who are committed to the highest standards of quality work. There are several key elements of such a policy:

- The required levels of technical and professional qualifications for positions in the agency are identified, and the agency adheres to these requirements in recruitment and professional development of staff. Position requirements take account of the different kinds of technical and other skills, such as supervisory skills, that are necessary for an agency to have a full range of qualified staff, including not only statisticians, but also experts in relevant subject-matter areas, data collection, processing, and dissemination processes, and management of complex, technical operations.
- Continuing technical education and training of staff, appropriate to the needs of their positions, are provided by sponsoring in-house training programs and providing opportunities for external education and training.
- Professional activities, such as publication in refereed journals and presentations at conferences, are encouraged and recognized. Participation in relevant statistical and other scientific associations is encouraged to promote interactions with academic researchers and other data users. Such participation is also a mechanism for openness about the data provided.
- Interaction with other professionals is increased through technical advisory committees, supervision of contract research and research consultants, fellowship programs of visiting researchers, exchange of staff with relevant statistical, policy, or research organizations, and opportunities for new assignments within the agency.
- Accomplishment is rewarded by appropriate recognition and by af-

fording opportunity for further professional development. The prestige and credibility of a statistical agency is enhanced by the professional visibility of its staff, which may include establishing high-level nonmanagement positions for highly qualified technical experts.

An effective statistical agency also has policies and practices to instill the highest possible commitment to professional ethics among its staff. When an agency comes under pressure to act against its principles—for example, if it is asked to disclose confidential information for an enforcement purpose or to support an inaccurate interpretation of its data—it must be able to rely on its staff to resist such actions as contrary to the ethical principles of their profession. An effective agency will refer its staff to such statements of professional practice as the guidelines published by the American Statistical Association and the International Statistical Institute on their Internet web sites (www.amstat.org/profession; www.cbs.nl/isi/ethics.htm), as well as to the agency's own statements about protection of confidentiality and similar matters. It will endeavor in other ways to ensure that its staff are fully cognizant of the ethics that must guide their actions in order for the agency to maintain its credibility as a source of objective, reliable information for use by all.

Coordination and Cooperation with Other Statistical Agencies

The U.S. federal statistical system consists of many agencies in different departments, each with its own mission. Nonetheless, statistical agencies do not and should not conduct their activities in isolation. An effective statistical agency will actively explore ways to work with other agencies to meet current information needs, for example, by seeking ways to integrate the designs of existing data systems to provide new or more useful data than a single system can provide. An effective agency will also be alert for occasions when it can provide technical assistance to other agencies—including not only other statistical agencies, but also program agencies in its department—as well as occasions on which it can receive such assistance in turn. Efforts to standardize definitions further contribute to effective coordination of statistical agency endeavors, as does the development of broad macro models, such as the system of national accounts. Initiatives for sharing data among statistical agencies (possibly including individual data and address lists when permitted by law and when sharing does not violate confidentiality promises) can also be helpful for such purposes as achieving greater efficiency in drawing samples or reducing duplication among statistical programs.

The responsibility for coordinating statistical work in the federal government is specifically assigned to the Office of Information and Regulatory Affairs (OIRA) in the Office of Management and Budget by the Paperwork Reduction Act (previously, the Federal Reports Act and the Budget and Accounting Procedures Act). Some functions are undertaken by OIRA desk officers; others, by the OMB Statistical Policy Office. Under the Paperwork Reduction Act, OIRA desk officers review proposed data collection instruments. The Statistical Policy Office, generally working with the assistance of interagency committees, reviews concepts of interest to more than one agency; issues standard classification systems (of industries, metropolitan areas, etc.) and oversees their periodic revision; consults with other parts of OMB on statistical budgets; and, by reviewing the statistical program of the government as a whole, identifies gaps in statistical data, programs that may be duplicative, and areas in which interagency cooperation might lead to greater efficiency and added utility of data. The Statistical Policy Office also is responsible for coordinating U.S. participation in international statistical activities.

The Statistical Policy Office encourages the use of administrative data for statistical purposes, when feasible, and works to establish common goals and norms on major statistical issues, such as confidentiality. It sponsors and heads the interagency Federal Committee on Statistical Methodology, which issues guidelines and recommendations on statistical issues common to a number of agencies (see Federal Committee on Statistical Methodology, 1978a-2000; see also www.fcsm.gov). It has encouraged the Committee on National Statistics at the National Academies to serve as an independent adviser and reviewer of federal statistical activities. The 1995 reauthorization of the Paperwork Reduction Act created the Interagency Council on Statistical Policy (ICSP), formalizing an arrangement whereby statistical agency heads participate with OMB to coordinate federal statistical activities. (See Box II-1 for a list of agencies represented on the ICSP.)

There are many forms of interagency cooperation and coordination. Some efforts are multilateral, some bilateral. Many result from common interests in specific subject areas, such as economic statistics, statistics on people with disabilities, or statistics on children or the elderly. (See U.S. Office of Management and Budget, 2000, for a description of several current interagency collaborative efforts.)

A common type of bilateral arrangement is the agreement of a program agency to provide administrative data to a statistical agency to be used as a sampling frame, a source of classification information, or a summary compilation to check (and possibly revise) preliminary sample results. The Bureau of Labor Statistics, for example, benchmarks its monthly establishment employment reports to data supplied by state employment security agencies. Such practices improve statistical estimates, reduce costs, and eliminate duplicate requests for information from the same respondents. In other cases, federal statistical agencies engage in cooperative data collection with state counterparts to let one collection system satisfy the needs of both. A number of such joint systems have been developed, notably by the Bureau of Labor Statistics, the National Agricultural Statistics Service, the National Center for Education Statistics, and the National Center for Health Statistics.

Another example of a joint arrangement is the case in which one statistical agency contracts with another to conduct a survey, compile special tabulations, or develop models. Such arrangements make use of the special skills of the supplying agency and facilitate use of common concepts and methods. The Bureau of the Census conducts many surveys for other agencies, as do the Bureau of Labor Statistics, the National Center for Health Statistics, and the National Agricultural Statistics Service. (See U.S. Office of Management and Budget, 2000, for a discussion of these and other reimbursable arrangements.)

The major federal statistical agencies are also concerned with international comparability of statistics. Under the overall guidance of OMB's Statistical Policy Office, they contribute to the deliberations of the United Nations Statistical Commission and other international organizations, participate in the development of international standard classifications and systems, and support educational activities that promote improved statistics in developing countries. Several statistical agencies run educational programs for government statisticians in developing countries. Some statistical agencies have had long-term cooperative relationships with international groups, for example, the Bureau of Labor Statistics with the International Labor Organization, the National Agricultural Statistics Service with the Food and Agriculture Organization, and the National Center for Health Statistics with the World Health Organization.

To be of most value, the efforts of statistical agencies to cooperate as partners with one another should involve the full range of their activities, including definitions, concepts, measurement methods, analytical tools, professional practice, dissemination modes, and disclosure limitation techniques. Such efforts should also extend to the development of data, especially for emerging policy issues (National Research Council, 1999a). In some cases, it may be not only more efficient, but also productive of needed new data for agencies to fully integrate the designs of existing data systems, such as when one survey provides the sampling frame for a related survey. In other instances, cooperative efforts may identify ways for agencies to improve their individual data systems so that they are more useful for a wide range of purposes.

Two of the more effective continuing cooperative efforts in this regard have been the Federal Interagency Forum on Aging-Related Statistics and the Federal Interagency Forum on Child and Family Statistics. The former was established in the mid-1980s by the National Institute on Aging, in cooperation with the National Center for Health Statistics and the Census Bureau. The forum's goals include coordinating the development and use of statistical data bases among federal agencies, identifying information gaps and data inconsistencies, and encouraging cross-national research and data collection for the aging population. The forum was reorganized in 1998 to include six new member agencies, and the reconfigured forum decided at its first meeting in March 1999 to focus on developing an indicators chart book, which was published the following year (Federal Interagency Forum on Aging-Related Statistics, 2000).

The Federal Interagency Forum on Child and Family Statistics was formalized in a 1994 executive order to foster coordination and collaboration in the collection and reporting of federal data on children and families. It includes many relevant statistical and program agencies. Its annual reports (e.g., Federal Interagency Forum on Child and Family Statistics, 2000) describe the condition of America's children, including changing population, family characteristics, and the context in which children are living and indicators of well-being in the areas of economic security, health, behavior, social environment, and education.

No single agency, whether a statistical or program agency, could have produced the forum reports alone. Working together in this way, federal statistical agencies contribute to data more relevant to policy concerns and to a stronger statistical system overall.

References

Duncan, Joseph W., and William C. Shelton

1978 *Revolution in United States Government Statistics, 1926-1976.* Office of Federal Statistical Policy and Standards. Washington, D.C.: U.S. Department of Commerce.

Federal Committee on Statistical Methodology

1978a *Report on Statistics for Allocation of Funds.* Statistical Policy Working Paper 1. (NTIS PB86-211521/AS) Washington, D.C.: U.S. Department of Commerce.

1978b *Report on Statistical Disclosure and Disclosure-Avoidance Techniques.* Statistical Policy Working Paper 2. (NTIS PB86-211539/AS) Washington, D.C.: U.S. Department of Commerce.

1978c *An Error Profile: Employment as Measured by the Current Population Survey.* Statistical Policy Working Paper 3. (NTIS PB86-214269/AS) Washington, D.C.: U.S. Department of Commerce.

1978d *Glossary of Nonsampling Error Terms: An Illustration of a Semantic Problem in Statistics.* Statistical Policy Working Paper 4. (NTIS PB86-211547/AS) Washington, D.C.: U.S. Department of Commerce.

1980a *Report on Exact and Statistical Matching Techniques.* Statistical Policy Working Paper 5. (NTIS PB86-215829/AS) Washington, D.C.: U.S. Department of Commerce.

1980b *Report on Statistical Uses of Administrative Records.* Statistical Policy Working Paper 6. (NTIS PB86-214285/AS) Washington, D.C.: U.S. Department of Commerce.

1982a *An Interagency Review of Time-Series Revision Policies.* Statistical Policy Working Paper 7. (NTIS PB86-232451/AS) Washington, D.C.: U.S. Office of Management and Budget.

1982b *Statistical Interagency Agreements.* Statistical Policy Working Paper 8. (NTIS PB86-230570/AS) Washington, D.C.: U.S. Office of Management and Budget.

1983a *Contracting for Surveys.* Statistical Policy Working Paper 9. (NTIS PB83-233148) Washington, D.C.: U.S. Office of Management and Budget.

1983b *Approaches to Developing Questionnaires.* Statistical Policy Working Paper 10. (NTIS PB84-105055) Washington, D.C.: U.S. Office of Management and Budget.

1984a *A Review of Industry Coding Systems.* Statistical Policy Working Paper 11. (NTIS PB84-135276) Washington, D.C.: U.S. Office of Management and Budget.

1984b *The Role of Telephone Data Collection in Federal Statistics.* Statistical Policy Working Paper 12. (NTIS PB85-105971) Washington, D.C.: U.S. Office of Management and Budget.

1986 *Federal Longitudinal Surveys.* Statistical Policy Working Paper 13. (NTIS PB86-139730) Washington, D.C.: U.S. Office of Management and Budget.

1987 *Workshop on Statistical Uses of Microcomputers in Federal Agencies.* Statistical Policy Working Paper 14. (NTIS PB87-166393) Washington, D.C.: U.S. Office of Management and Budget.

1988 *Quality in Establishment Surveys.* Statistical Policy Working Paper 15. (NTIS PB88-232921) Washington, D.C.: U.S. Office of Management and Budget.

1990a *A Comparative Study of Reporting Units in Selected Employer Data Systems.* Statistical Policy Working Paper 16. (NTIS PB90-205238) Washington, D.C.: U.S. Office of Management and Budget.

1990b *Survey Coverage.* Statistical Policy Working Paper 17. (NTIS PB90-205246) Washington, D.C.: U.S. Office of Management and Budget.

1990c *Data Editing in Federal Statistical Agencies.* Statistical Policy Working Paper 18. (NTIS PB90-205253) Washington, D.C.: U.S. Office of Management and Budget.

1990d *Computer Assisted Survey Information.* Statistical Policy Working Paper 19. (NTIS PB90-205261) Washington, D.C.: U.S. Office of Management and Budget.

1991 *Seminar on Quality of Federal Data.* Statistical Policy Working Paper 20. (NTIS PB91-142414) Washington, D.C.: U.S. Office of Management and Budget.

1993 *Indirect Estimators in Federal Programs.* Statistical Policy Working Paper 21. (NTIS PB93-209294) Washington, D.C.: U.S. Office of Management and Budget.

1994 *Report on Statistical Disclosure Limitation Methodology.* Statistical Policy Working Paper 22. (NTIS PB94-165305) Washington, D.C.: U.S. Office of Management and Budget.

1995a *Seminar on New Directions in Statistical Methodology.* Statistical Policy Working Paper 23. (NTIS PB95-182978) Washington, D.C.: U.S. Office of Management and Budget.

1995b *Electronic Dissemination of Statistical Data.* Statistical Policy Working Paper 24. (NTIS PB96-121629) Washington, D.C.: U.S. Office of Management and Budget.

1996 *Data Editing Workshop and Exposition.* Statistical Policy Working Paper 25. (NTIS PB97-104624) Washington, D.C.: U.S. Office of Management and Budget.

1997 *Seminar on Statistical Methodology in the Public Service.* Statistical Policy Working Paper 26. (NTIS PB97-162580) Washington, D.C.: U.S. Office of Management and Budget.

1998 *Training for the Future: Addressing Tomorrow's Survey Tasks.* Statistical Policy

Working Paper 27. (NTIS PB99-102576) Washington, D.C.: U.S. Office of Management and Budget.

1999a *Seminar on Interagency Coordination and Cooperation.* Statistical Policy Working Paper 28. (NTIS PB99-132029) Washington, D.C.: U.S. Office of Management and Budget.

1999b *Federal Committee on Statistical Methodology Research Conference (Conference Papers).* Statistical Policy Working Paper 29. (NTIS PB99-166795) Washington, D.C.: U.S. Office of Management and Budget.

2000 *1999 Federal Committee on Statistical Methodology Research Conference: Complete Proceedings.* Statistical Policy Working Paper 30. (NTIS PB2000-105886) Washington, D.C.: U.S. Office of Management and Budget.

Federal Interagency Forum on Aging-Related Statistics

2000 *Older Americans 2000: Key Indicators of Well-Being.* Available online (http://www.agingstats.gov).

Federal Interagency Forum on Child and Family Statistics

2000 *America's Children: Key National Indicators of Well-Being, 2000.* Washington, D.C.: U.S. Government Printing Office.

Martin, Margaret E.

1981 Statistical practice in bureaucracies. *Journal of the American Statistical Association* 76(373):1-8.

National Research Council

1976 *Setting Statistical Priorities: Report of the Panel on Methodology for Statistical Priorities of the Committee on National Statistics.* Assembly of Mathematical and Physical Sciences, Committee on National Statistics. Washington, D.C.: National Academy Press.

1984 *Cognitive Aspects of Survey Methodology: Building a Bridge Between Disciplines.* Report of the Advanced Research Seminar on Cognitive Aspects of Survey Methodology, Thomas B. Jabine, Miron L. Straf, Judith M. Tanur, and Roger Tourangeau, eds. Committee on National Statistics. Washington, D.C.: National Academy Press.

1985a *Immigration Statistics: A Story of Neglect.* Panel on Immigration Statistics, Daniel B. Levine, Kenneth Hill, and Robert Warren, eds. Committee on National Statistics. Washington, D.C.: National Academy Press.

1985b *Sharing Research Data.* Stephen E. Fienberg, Margaret E. Martin, and Miron L. Straf, eds. Committee on National Statistics. Washington, D.C.: National Academy Press.

1986 *Creating a Center for Education Statistics: A Time for Action.* Panel to Evaluate the National Center for Education Statistics, Daniel B. Levine, ed. Committee on National Statistics. Washington, D.C.: National Academy Press.

1991a *Improving Information for Social Policy Decisions: The Uses of Microsimulation Modeling, Volume I: Review and Recommendations.* Panel to Evaluate Microsimulation Models for Social Welfare Programs, Constance F. Citro and Eric A. Hanushek, eds. Committee on National Statistics. Washington, D.C.: National Academy Press.

1991b *Improving Information for Social Policy Decisions: The Uses of Microsimulation Modeling, Volume II: Technical Papers.* Panel to Evaluate Microsimulation Models for Social Welfare Programs, Constance F. Citro and Eric A. Hanushek, eds. Committee on National Statistics. Washington, D.C.: National Academy Press.

1992 *Principles and Practices for a Federal Statistical Agency.* Margaret E. Martin and Miron L. Straf, eds. Committee on National Statistics. Washington, D.C.: National Academy Press.

1993a *The Future of the Survey of Income and Program Participation.* Panel to Evaluate the Survey of Income and Program Participation, Constance F. Citro and Graham Kalton, eds. Committee on National Statistics. Washington, D.C.: National Academy Press.

1993b *Private Lives and Public Policies: Confidentiality and Accessibility of Government Statistics.* Panel on Confidentiality and Data Access, George T. Duncan, Thomas B. Jabine, and Virginia A. de Wolf, eds. Committee on National Statistics. Washington, D.C.: National Academy Press.

1995 *Modernizing the U.S. Census.* Panel on Census Requirements in the Year 2000 and Beyond. Barry Edmonston and Charles Schultze, eds. Committee on National Statistics. Washington, D.C.: National Academy Press.

1997a *Assessing Policies for Retirement Income: Needs for Data, Research, and Models.* Panel on Retirement Income Modeling, Constance F. Citro and Eric A. Hanushek, eds. Committee on National Statistics. Washington, D.C.: National Academy Press.

1997b *The Bureau of Transportation Statistics: Priorities for the Future.* Panel on Statistical Programs and Practices of the Bureau of Transportation Statistics, Constance F. Citro and Janet L. Norwood, eds. Committee on National Statistics and Transportation Research Board. Washington, D.C.: National Academy Press.

1998a *Longitudinal Surveys of Children.* Committee on National Statistics and Board on Children, Youth, and Families, Kirsten K. West, Robert M. Hauser, and Terri M. Scanlan, eds. Washington, D.C.: National Academy Press.

1998b *Statistics, Testing, and Defense Acquisition: New Approaches and Methodological Improvements.* Panel on Statistical Methods for Testing and Evaluating Defense Systems, Michael L. Cohen, John E. Rolph, and Duane L. Steffey, eds. Committee on National Statistics. Washington, D.C.: National Academy Press.

1999a *Health Performance Measurement in the Public Sector: Principles and Policies for Implementing an Information Network.* Panel on Performance Measures and Data for Public Health Performance Partnership Grants, Edward B. Perrin, Jane S. Durch, and Susan M. Skillman, eds. Committee on National Statistics. Washington, D.C.: National Academy Press.

1999b *Sowing Seeds of Change: Informing Public Policy in the Economic Research Service of USDA.* Panel to Study the Research Program of the Economic Research Service, John F. Geweke, James T. Bonnen, Andrew A. White, and Jeffrey J. Koshel, eds. Committee on National Statistics. Washington, D.C.: National Academy Press.

2000 *Improving Access to and Confidentiality of Research Data: Report of a Workshop.* Christopher Mackie and Norman Bradburn, eds. Committee on National Statistics. Washington, D.C.: National Academy Press.

National Research Council and Institute of Medicine

1992 *Toward a National Health Care Survey: A Data System for the 21st Century.* Panel on the National Health Care Survey, Gooloo S. Wunderlich, ed. Committee on

National Statistics and Division of Health Care Services. Washington, D.C.: National Academy Press.

Norwood, Janet L.

1975 Should those who produce statistics analyze them? How far should analysis go? An American view. *Bulletin of the International Statistical Institute*, Proceedings of the 40th session, 46:420-432.

1995 *Organizing to Count: Change in the Federal Statistical System.* Washington, D.C.: The Urban Institute Press.

President's Commission on Federal Statistics

1971 *Federal Statistics. Vol. I.* Washington, D.C.: U.S. Government Printing Office.

Privacy Protection Study Commission

1977 *Personal Privacy in an Information Society.* Washington, D.C: U.S. Government Printing Office.

Ryten, Jacob

1990 Statistical organization criteria for inter-country comparisons and their application to Canada. *Journal of Official Statistics* 6(3):319-332.

Triplett, Jack

1991 The federal statistical system's response to emerging data needs. *Journal of Economic and Social Measurement* 17(3,4):155-201.

United Nations Statistical Commission

1994 Fundamental Principles of Official Statistics. In Official Records of the Economic and Social Council, 1994, Supplement No. 9.

United Nations Statistical Commission and Economic Commission for Europe

1991 Draft of a Resolution on fundamental principles of official statistics in the Economic Commission for Europe. In *Report of the Thirty-Ninth Plenary Session,* June 17-21, 1991.

U.S. General Accounting Office

1995 *Statistical Agencies: Adherence to Guidelines and Coordination of Budgets.* Washington, D.C.

U.S. Office of Management and Budget

1985 Statistical policy directive on compilation, release, and evaluation of principal federal economic indicators. *Federal Register* 50(186):932-934.

1997 Order providing for the confidentiality of statistical information. *Federal Register* 62(124):35044-35050.

2000 *Statistical Programs of the United States Government, Fiscal Year 2001.* Statistical Policy Office, Office of Information and Regulatory Affairs, U.S. Office of Management and Budget. Washington, D.C.: U.S. Government Printing Office.

Appendix A
Fundamental Principles of Official Statistics of the Statistical Commission of the United Nations

[Excerpted from the report of the Statistical Commission on its Special Session, held in New York 11-15 April 1994. Official Records of the Economic and Social Council, 1994, Supplement No. 9.]

Action taken by the Commission

59. The Commission adopted the fundamental principles of official statistics as set out in ECE decision C (47), but incorporating a revised preamble. The preamble and principles, as adopted, are set out below:

FUNDAMENTAL PRINCIPLES OF OFFICIAL STATISTICS

The Statistical Commission,

Bearing in mind that official statistical information is an essential basis for development in the economic, demographic, social and environmental fields and for mutual knowledge and trade among the States and peoples of the world,

Bearing in mind that the essential trust of the public in official statistical information depends to a large extent on respect for the fundamental values and principles which are the basis of any society which seeks to understand itself and to respect the rights of its members,

Bearing in mind that the quality of official statistics, and thus the quality of the information available to the Government, the economy and the public depends largely on the cooperation of citizens, enterprises, and other respondents in providing appropriate and reliable data needed for necessary statistical compilations and on the cooperation between users and producers of statistics in order to meet users' needs,

Recalling the efforts of governmental and non-governmental organizations active in statistics to establish standards and concepts to allow comparisons among countries,

Recalling also the International Statistical Institute Declaration of Professional Ethics,

Having expressed the opinion that resolution C (47), adopted by the Economic Commission for Europe on 15 April 1992, is of universal significance,

Noting that, at its eighth session, held at Bangkok in November 1993, the Working Group of Statistical Experts, assigned by the Committee on Statistics of the Economic and Social Commission for Asia and the Pacific to examine the Fundamental Principles, had agreed in principle to the ECE version and had emphasized that those principles were applicable to all nations,

Noting also that, at its eighth session, held at Addis Ababa in March 1994, the Joint Conference of African Planners, Statisticians and Demographers, considered that the Fundamental Principles of Official Statistics are of universal significance,

Adopts the present principles of official statistics:

1. Official statistics provide an indispensable element in the information system of a democratic society, serving the Government, the economy and the public with data about the economic, demographic, social and environmental situation. To this end, official statistics that meet the test of practical utility are to be compiled and made available on an impartial basis by official statistical agencies to honour citizens' entitlement to public information.

2. To retain trust in official statistics, the statistical agencies need to decide according to strictly professional considerations, including scientific principles and professional ethics, on the methods and procedures for the collection, processing, storage and presentation of statistical data.

3. To facilitate a correct interpretation of the data, the statistical agencies are to present information according to scientific standards on the sources, methods and procedures of the statistics.

4. The statistical agencies are entitled to comment on erroneous interpretation and misuse of statistics.

5. Data for statistical purposes may be drawn from all types of sources, be they statistical surveys or administrative records. Statistical agencies are to choose the source with regard to quality, timeliness, costs and the burden on respondents.

6. Individual data collected by statistical agencies for statistical compilation, whether they refer to natural or legal persons, are to be strictly confidential and used exclusively for statistical purposes.

7. The laws, regulations and measures under which the statistical systems operate are to be made public.

8. Coordination among statistical agencies within countries is essential to achieve consistency and efficiency in the statistical system.

9. The use by statistical agencies in each country of international concepts, classifications and methods promotes the consistency and efficiency of statistical systems at all official levels.

10. Bilateral and multilateral cooperation in statistics contributes to the improvement of systems of official statistics in all countries.

Appendix B
Selected Federal Statistical World Wide Web Sites, May 2000

Executive Office of the President
- Federal Statistics Briefing Rooms
  - Economic Statistics (ESBR): www.whitehouse.gov/fsbr/esbr.html
  - Social Statistics (SSBR): www.whitehouse.gov/fsbr/ssbr.html
- Office of Management and Budget (OMB): www.whitehouse.gov/OMB
  - Federal Committee on Statistical Methodology: www.fcsm.gov
  - Federal Interagency Council on Statistical Policy, Federal Statistics: www.fedstats.gov
  - Federal Interagency Forum on Aging-Related Statistics: www.agingstats.gov
  - Federal Interagency Forum on Child and Family Statistics: childstats.gov

Consumer Product Safety Commission (CPSC): www.cpsc.gov

Department of Agriculture (USDA): www.usda.gov
- Agricultural Research Service (ARS): www.ars.usda.gov

Note: This list is adapted from a listing in U.S. Office of Management and Budget (2000) and includes federal statistical agencies, as well as other federal agencies that produce statistical information.

Economic Research Service (ERS): www.ers.usda.gov
Food and Nutrition Service (FNS): www.fns.usda.gov/fns
Foreign Agricultural Service (FAS): www.fas.usda.gov
Forest Service (FS): www.fs.fed.us
National Agricultural Statistics Service (NASS): www.usda.gov/nass
Natural Resources Conservation Service (NRCS): www.nrcs.usda.gov

Department of Commerce: www.doc.gov
Bureau of Economic Analysis (BEA): www.bea.doc.gov
Census Bureau: www.census.gov
Economics and Statistics Administration (ESA): www.esa.doc.gov
International Trade Administration (ITA): www.ita.doc.gov
National Marine Fisheries Service (NMFS): www.nmfs.gov
National Oceanic and Atmospheric Administration (NOAA): www.noaa.gov

Department of Defense: www.defenselink.mil
Army Corps of Engineers (CORPS): www.environmental.usace.army.mil
Defense Manpower Data Center (DMDC): www.dmdc.osd.mil
Directorate for Information Operations and Reports (DIOR): web1.whs.osd.mil/diorhome.htm

Department of Education: www.ed.gov
National Center for Education Statistics (NCES): nces.ed.gov

Department of Energy: www.doe.gov
Energy Information Administration (EIA): www.eia.doe.gov
Office of Environment, Safety and Health (EH): www.eh.doe.gov

Department of Health and Human Services: www.dhhs.gov
Administration for Children and Families (ACF): www.acf.dhhs.gov
Agency for Health Care Research and Quality (AHRQ): www.ahcpr.gov
Agency for Toxic Substances and Disease Registry (ATSDR): www.atsdr.cde.gov
Centers for Disease Control and Prevention (CDC): www.cdc.gov
Health Care Financing Administration (HCFA): www.hcfa.gov

Health Resources and Services Administration (HRSA): www.hrsa.dhhs.gov
Indian Health Service (IHS): www.ihs.gov
National Center for Health Statistics (NCHS): www.cdc.gov/nchs
National Institute on Aging (NIA): www.nih.gov/nia
National Institutes of Health (NIH): www.nih.gov
Office of Population Affairs (OPA): www.dhhs.gov/progorg/opa
Office of the Assistant Secretary for Planning and Evaluation (ASPE): aspe.os.dhhs.gov
Substance Abuse and Mental Health Services Administration (SAMHSA): www.samhsa.gov

Department of Housing and Urban Development: www.hud.gov
Office of the Assistant Secretary for Policy Development and Research (PD&R): www.huduser.org

Department of the Interior: www.doi.gov
Bureau of Reclamation: www.usbr.gov
Minerals Management Service (MMS): www.mms.gov
National Park Service (NPS): www.nps.gov
United States Fish and Wildlife Service (FWS): www.fws.gov
United States Geological Survey (USGS): www.usgs.gov

Department of Justice: www.usdoj.gov
Bureau of Justice Statistics (BJS): www.ojp.usdoj.gov/bjs
Bureau of Prisons (BoP): www.bop.gov
Drug Enforcement Administration (DEA): www.usdoj.gov/dea
Federal Bureau of Investigation (FBI): www.fbi.gov
Immigration and Naturalization Service (INS): www.ins.usdoj.gov

Department of Labor: www.dol.gov
Bureau of Labor Statistics (BLS): www.bls.gov
Employment and Training Administration (ETA): www.doleta.gov
Mine Safety and Health Administration (MSHA): www.msha.gov
Occupational Safety and Health Administration (OSHA): www.osha.gov
Office of the Assistant Secretary for Policy (OASP): www.dol.gov/dol/asp

Pension and Welfare Benefits Administration (PWBA): www.dol.gov/dol/pwba

Department of State: www.state.gov
Agency for International Development (AID): www.usaid.gov

Department of Transportation: www.dot.gov
Bureau of Transportation Statistics (BTS): www.bts.gov
Federal Aviation Administration (FAA): www.faa.gov
Federal Highway Administration (FHWA): www.fhwa.dot.gov
Federal Railroad Administration (FRA): www.fra.dot.gov
Federal Transit Administration (FTA): www.fta.dot.gov
Maritime Administration (MARAD): www.marad.dot.gov
National Highway Traffic Safety Administration (NHTSA): www.nhtsa.dot.gov/people/ncsa

Department of the Treasury: www.ustreas.gov
Internal Revenue Service (IRS): www.irs.ustreas.gov
Statistics of Income (SOI): www.irs.ustreas.gov/tax_stats/index.html
United States Customs Service: www.customs.treas.gov

Department of Veterans Affairs (VA): www.va.gov

Environmental Protection Agency (EPA): www.epa.gov
Office of Environmental Information: www.epa.gov/oei

Federal Emergency Management Agency (FEMA): www.fema.gov

National Aeronautics and Space Administration (NASA): www.nasa.gov

National Science Foundation: www.nsf.gov/sbe/srs
Directorate for Social, Behavioral, and Economic Sciences: www.nsf.gov/sbe
Methodology, Measurement, and Statistics Program: www.nsf.gov/sbe/ses/mms/start.htm
Science Resources Studies Division: www.nsf.gov/sbe/srs/stats.htm

Small Business Administration (SBA): www.sba.gov/advo/stats

Social Security Administration (SSA): www.ssa.gov
Office of Research, Evaluation, and Statistics: www.ssa.gov/statistics/ores_home.html

INDEX OF FEDERAL STATISTICAL SITES LISTED

Energy Information Administration—see Department of Energy
Environmental Protection Agency—see Environmental Protection Agency
Federal Aviation Administration—see Department of Transportation
Federal Bureau of Investigation—see Department of Justice
Federal Committee on Statistical Methodology—see Executive Office of the President, Office of Management and Budget
Federal Emergency Management Agency—see Federal Emergency Management Agency
Federal Highway Administration—see Department of Transportation
Federal Interagency Council on Statistical Policy, Federal Statistics—see Executive Office of the President, Office of Management and Budget
Federal Interagency Forum on Aging-Related Statistics—see Executive Office of the President, Office of Management and Budget
Federal Interagency Forum on Child and Family Statistics—see Executive Office of the President, Office of Management and Budget
Federal Railroad Administration—see Department of Transportation
Federal Statistics Briefing Rooms—see Executive Office of the President
Federal Transit Administration—see Department of Transportation
Food and Nutrition Service—see Department of Agriculture
Foreign Agricultural Service—see Department of Agriculture
Forest Service—see Department of Agriculture
Health Care Financing Administration—see Department of Health and Human Services
Health Resources and Services Administration—see Department of Health and Human Services Immigration and Naturalization Service—see Department of Justice
Indian Health Service—see Department of Health and Human Services
Internal Revenue Service—see Department of the Treasury
International Trade Administration—see Department of Commerce
Maritime Administration—see Department of Transportation
Methodology, Measurement, and Statistics Program—see National Science Foundation, Directorate for Social, Behavioral, and Economic Sciences
Mine Safety and Health Administration—see Department of Labor
Minerals Management Service—see Department of the Interior
National Aeronautics and Space Administration—see National Aeronautics and Space Administration
National Agricultural Statistics Service—see Department of Agriculture

National Center for Education Statistics—see Department of Education
National Center for Health Statistics—see Department of Health and Human Services
National Highway Traffic Safety Administration—see Department of Transportation
National Institute on Aging—see Department of Health and Human Services
National Institutes of Health—see Department of Health and Human Services
National Marine Fisheries Service—see Department of Commerce
National Oceanic and Atmospheric Administration—see Department of Commerce
National Park Service—see Department of the Interior
National Science Foundation—see National Science Foundation
Natural Resources Conservation Service—see Department of Agriculture
Occupational Safety and Health Administration—see Department of Labor
Office of Environmental Information—see Environmental Protection Agency
Office of the Assistant Secretary for Planning and Evaluation—see Department of Health and Human Services
Office of the Assistant Secretary for Policy—see Department of Labor
Office of the Assistant Secretary for Policy Development and Research—see Department of Housing and Urban Development
Office of Environment, Safety and Health—see Department of Energy
Office of Environmental Information—see Environmental Protection Agency
Office of Management and Budget—see Executive Office of the President
Office of Population Affairs—see Department of Health and Human Services
Office of Research, Evaluation, and Statistics—see Social Security Administration
Pension and Welfare Benefits Administration—see Department of Labor
Science Resources Studies Division—see National Science Foundation, Directorate for Social, Behavioral, and Economic Sciences
Small Business Administration—see Small Business Administration
Social Security Administration—see Social Security Administration
Social Statistics Briefing Room—see Executive Office of the President, Federal Statistics Briefing Rooms

Statistics of Income Division—see Department of the Treasury
Substance Abuse and Mental Health Services Administration—see Department of Health and Human Services
United States Customs Service—see Department of the Treasury
United States Fish and Wildlife Service—see Department of the Interior
United States Geological Survey—see Department of the Interior